European Tours of Duty

Vickey Joins the Military Series

Book 4

European Tours of Duty

Book Four

Table of Contents

South-Central Germany-Part I

First Tour, First Stop

First Stop
Frankfurt, GE
Replacement BN

As the aircraft landed in Germany, all military personnel disembarked and formed up; structure is fundamental in the military. We marched to a large room designated for our in-processing.

At that time, anyone holding the rank of Private First Class or Specialist was selected for interviews for Bus Commanders at the reception station. From ten candidates, only two of us were chosen to remain until Monday for the interview process for the position.

We learned that the position would grant us a certificate of merit, a recommendation, and our choice of duty station, so we accepted the role. Wow what an incredible way to begin our tour of duty in Germany.

I was in Germany, embarking on a life of travel. My European adventure was off to a great start. My new role as a Bus Commander involved riding the bus from the Replacement Battalion to various destinations in Germany, transporting incoming personnel to their respective units.

My primary duty was to ensure that US military personnel arriving in the country reached their designated duty stations. Receiving orders to Germany, a foreign land unfamiliar to many, could be an intimidating experience. We were there to alleviate much of that anxiety.

One month into my tour of duty at the Replacement Battalion, we were visited by a very high-ranking official from the Joint Chiefs of Staff, General Colin Powell. I had the opportunity to meet, speak with, and shake hands with General Powell, the highest military figure in the US.

Shaking hands with General Colin Powell, JCoS!

With only seven months of service under my belt, I had already met Vice President Dan Quayle, Sergeant Major of the Army Julius W. Gates, and General Colin Powell of the Joint Chiefs of Staff. Despite being in formation, I broke ranks to take photos, determined not to miss another significant opportunity—after all, I was a former journalist.

As my second month at the Replacement Battalion began, operations continued smoothly. We kept transporting personnel to their new duty stations, but we also looked for ways to break the routine. We all volunteered for a Junior Olympic event being held in the area.

While stationed at the Reception Battalion, I explored several career advancement options within the military. I considered applying for Officer Candidate School (OCS) but was told I was too old. I thought about becoming a Warrant Officer (WO), yet I lacked the necessary rank. I looked into changing my military occupational specialty, but the Army was not supportive of that change. Finally, I considered working in broadcast news, leveraging my second degree in that field, but that too seemed unpromising as the Army insisted I stick to the job they had assigned me.

Despite the challenges, I visited the local broadcast station, Armed Forces Network (AFN), to see if I could work there in my spare time. They agreed, but everything depended on my availability. My hope was that when it came time to choose my permanent duty station, I could secure a "letter of acceptance" from AFN, which would allow me to work there officially.

As my second month at the battalion drew to a close, my attempts to explore different career paths within the Army had not been successful. I was fairly certain I had failed the OCS exam. Additionally, I lacked a car for transportation, was experiencing unresolved issues with my pay, and was unsure of the cause. My situation was becoming increasingly gloomy.

With just a few dollars left in my bank account, I urgently needed to figure out why I wasn't receiving my pay. Amidst the series of setbacks, I decided to channel my efforts into studying for the sergeant's board, thinking that if nothing else, I could at least aim for a promotion.

Now entering my third and final month at the Replacement Battalion, we continued to make the most of our down time. We embraced the adventure by taking the train to visit other cities across the country, turning our challenges into opportunities to explore.

The train system was always on time!

We also enjoyed our social time.

Despite experiencing some down days, once my pay situation was sorted out, things began to look up. It was then that I resolved to channel my efforts into completing my OCS packet, passing my European driving test, and securing employment at the AFN station.

Upon the arrival of my OCS exam results, I was taken aback by the outcome - I had passed. The disbelief stemmed from my lack of familiarity with the exam's content. Nonetheless, with the hardest part now behind me, my focus shifted to gathering the remaining materials required to finalize the packet.

After failing my first attempt at the driving test, I swiftly made arrangements for a retest. With my time at the Replacement Battalion drawing to a close, I eagerly awaited the arrival of my replacement to commence training. Simultaneously, I persisted in my search for a permanent duty station following my departure from the Battalion.

As my tenure as a Bus Commander reached its 90-day mark, my replacement was appointed. Subsequently, I successfully passed my driver's test on the second try, granting me the authorization to drive in Europe. Furthermore, I received orders assigning me to my inaugural permanent duty station in South-Central Germany, a development that thrilled me, particularly as I still harbored aspirations to work at AFN in Frankfurt.

Upon my arrival at my permanent duty station, I commenced the in-processing procedure into the unit. I was promptly assigned temporary living quarters and informed of my new role as a squad leader.

Welcome, Specialist Robinson, to your first European Tour!

South-Central Germany

First Tour, Second Stop

Second Stop
Permanent Duty

My tour of duty at my permanent duty station commenced at yet another reception area, as most days during the journey did, starting at "oh-dark-thirty."

At reception, it was the place where new soldiers passed through to take care of the processes needed to integrate into a new unit, and in that case, that included learning the country's language.

The days were filled with processes, while the nights exuded tranquility. During these nights, I could lie in bed with the window slightly ajar, the night air gently cooling the room, and the moon casting a bright glow through the window. Everything seemed calm and peaceful, devoid of any hint of the strict military existence.

As with all previous stops on the military journey, there were requirements to pass exams before being permitted to move on. The exams were successfully completed, followed by a graduation ceremony and the receipt of a certificate of completion. Subsequently,

I proceeded to my permanent duty station. Just like all other living quarters on the journey, my new home - half of a small room - required a much-needed cleaning and some materials to make it feel like home.

No Army olive drab wool blanket for me!

Following a thorough cleaning and the acquisition of a few essentials, my new quarters became habitable. Opting for personal comfort, I ventured out to purchase bedding, as I preferred not to sleep on the Army-issued linens.

Additionally, I invested in a television, a video cassette recorder, lamps, and various other items to ensure that the Army's olive drab green items wouldn't overshadow my existence. The room reached its full potential with the addition of wall fixtures, proudly displaying degrees and military certificates, symbolizing the amalgamation of my civilian and military journeys.

The half-a-room would be my home for two years!

Before embarking on my new job, I made every effort to put the final touches on my new living quarters. Successfully blending the Army furnishings with my civilian belongings, I curated a comfortable and inviting living space.

Arrival First Unit

Hit the ground running.

The anticipation was finally over as I arrived at my first permanent duty station. The first day on the job unfolded much like any other - filled with the uncertainty of new beginnings and the endeavor to grasp the ropes. However, a few occurrences deviated from the typical first-day scenario. The pace of the day was unexpectedly swift, and I found myself occupied from dawn to dusk, a departure from the usual leisurely pace associated with a new job.

The department to which I was assigned was understaffed, leaving me feeling like I had been thrown into the water without a paddle. That marked the commencement of numerous on-the-job training scenarios that would follow.

To unwind and destress at the end of the day, I headed to the gym. That visit also served as preparation for an upcoming PT test and an opportunity to shed a few pounds before a mandatory weigh-in.

The following day introduced me to something new: Sergeant's Time. Though unfamiliar with the event's details, I anticipated its significance, especially knowing that the Sergeant Major of the Army would be present to observe.

Later, I learned that Sergeant's Time was a weekly, five-hour period dedicated to sergeants training their squad members on various tasks, aimed at honing their combat skills. We drove out to a wooded area and spent about 20 minutes setting up equipment. The training focused on sending communication signals and conducting driver's training for military vehicles. Upon completion, we returned to the office to organize our department.

But it's not just work!

What the military brochures failed to mention was that upon arrival at our duty stations, the focus wasn't solely on performing our assigned tasks; it also involved continuous training. There was always some form of training taking place, whether in preparation for various scenarios or to stay up-to-date on essential skills.

Physical training tests occurred every six months, while requalifying on assigned weapons was an annual requirement or as frequently as necessary. Additionally, gas chamber events were conducted annually, and there was a perpetual need to maintain a fighting weight, requiring constant monitoring.

There were moments when we could set aside the military obligations and simply relish our time off, immersing ourselves in the rich culture of the country.

Oktoberfest 1991!

My first Oktoberfest experience in Germany commenced bright and early at 4:30 am. Equipped with a small carry bag containing food, headphones, and a jacket, I made my way to catch the bus arranged by the unit. Departing at 6:20 am, our bus journeyed towards Munich, Germany, taking approximately six hours to reach Oktoberfest.

Upon arrival, we were set free to revel in the festivities. It was during that celebration that I encountered my first authentic German custom. Alongside two sergeants, I indulged in beer and relished the German sausage. Initially, I likened the event to a grand carnival back home, but it boasted colossal pretzels and oversized glasses of beer, unlike anything I had ever seen before. It was an immensely enjoyable experience.

The workweek kicked off with its customary brisk pace, which was manageable at first. However, residing in the same building where I worked started to pose challenges. The proximity meant there was always an expectation that I could be called in early or required to work late. After enduring a month of constant demands for extra hours and work, I began to feel overwhelmed and undervalued. Nevertheless, I persevered through the strain, holding onto the hope of a forthcoming promotion.

While awaiting the promotion, I decided to lease a five-speed Ford Fiesta to enhance my leisure time. The lease consumed all my available funds, but the newfound freedom it provided was worth every penny. An interesting twist was that I didn't know how to drive a stick shift, so I had yet another skill to acquire.

As time passed and my day-to-day responsibilities became more manageable, I reflected on my journey in the Army. I realized that joining the Army had allowed me to fulfill some long-held desires, such as experiencing dorm life, traveling, and even pledging a sorority. Joining the Army had provided me with opportunities to tick off items from my bucket list.

The first unit field event.

With just about a month and a half at my first permanent duty station, we embarked on our first field deployment. It became apparent early on that field exercises would be a regular occurrence throughout my military career.

That field experience proved to be an improvement over my initial encounter. We had cots for our sleeping bags, which significantly enhanced the comfort and quality of sleep.

The exercise marked the beginning of many field excursions to follow. It involved intensive training and re-qualification on various military skills. However, amidst the rigorous training, there was a unique opportunity for adventure. After completing the training and qualification sessions, I had the chance to do something I had never considered before: rappel off a mountain.

Upon receiving orders from the commander to attend the rappel course, where we were expected to descend down the side of a mountain using a rope, my initial instinct was to find a way to avoid the course altogether. However, I eventually decided to at least observe the event.

Arriving at the site, I couldn't help but think, "This doesn't look too difficult, I'll give it a try." However, before attempting the feat, we had to first climb to the top of the mountain. The ascent left me a bit breathless, but I remained undaunted. It wasn't until I witnessed the first individual take the plunge, visibly gripped with fear, that doubts began to creep into my mind about rappelling off the side of that mountain.

As my turn approached to descend the mountain, armed with two ropes and relying on the trust of my battle buddies, I summoned my courage and geared up.

Starting my descent down the 90-foot-high mountain backwards, suspended from the two ropes, my initial moments were a bit unsteady. However, once I gained momentum, it felt strangely familiar, as if I had done it countless times before. Contrary to my expectations, it wasn't as daunting as it appeared. In fact, it was surprisingly easy, and I found myself enjoying the experience immensely. I even opted to repeat the descent a couple more times before the field event concluded. It was an activity that had never crossed my mind before, but it turned out to be an exhilarating and unforgettable experience.

As I returned to the campsite, the setting sun cast a serene atmosphere, accompanied by a cool evening breeze. It was a peaceful moment to wrap up what had been a good and exciting day.

With the conclusion of the field event, I couldn't deny that I had enjoyed it, but the prospect of heading home was the highlight. As days passed, my one-year mark in the military was swiftly approaching. I had envisioned doing something special to commemorate the occasion, but the day unfolded with the usual routine of work.

As my days carried on, another unexpected first emerged: undergoing a urinalysis. The day didn't commence with the typical mundane routine; instead, it began abruptly at 3:30 am, with sergeants hurriedly rushing down the hallway, shouting "get up, get dressed, and go to the theater."

My initial thought was, "Who are these crazy people?" Despite my confusion, I reluctantly dragged myself out of bed, hastily dressed, and made my way to the theater as instructed, unsure of what awaited me.

Upon arrival, we were informed that we were prohibited from using the bathroom at all, whether to relieve ourselves or even to brush our teeth. It soon became apparent that we were about to undergo a urinalysis test, with the added surprise that someone would be closely monitoring our every move.

Early in my military career, I quickly learned that privacy was a luxury, and newcomers could expect little of it. So, during the urinalysis, when someone closely monitored the event, I wasn't completely taken aback. It was a bit unnerving, but as with all things military-related, you learn to adapt and get through it.

After that unusual event, the first of many that would occur throughout my military journey, we all resumed our normal duty day.

Always learning something new.

Another first in my military journey was learning combat lifesaver skills, some of which were entirely new to me, like administering intravenous fluids or an IV. Surprisingly, it was an exciting and less daunting task than I had anticipated. Upon successful completion of the training, I attained certification as a combat lifesaver, equipped to provide basic, immediate care to personnel in a field environment.

With a little over a year in the military, I had already encountered and witnessed things in the Army that I never would have experienced within my comfort zone. Even if my tenure in the Army were to last only two years, the events I experienced would have been worthwhile. And the adventures were far from over; there was an upcoming Christmas event that promised to bring together the singing talents of both US and German soldiers.

Approaching my second Christmas in the Army, and my first as a full-fledged soldier, the unit was abuzz with the holiday spirit. Joining the unit choir, we commenced practicing for our upcoming performance alongside our German counterparts' choir.

After rigorous rehearsal sessions, we boarded a bus bound for the cultural event venue. The journey lasted about two-and-a-half hours, filled with excitement and anticipation. Upon arrival, we changed into our formal uniforms in preparation for the performance.

The program commenced punctually, and as we sat at our designated tables, we engaged in friendly exchanges with our German counterparts. As the moment for our performance drew near, I felt a tinge of nervousness, but it dissipated once we took the stage. If I may say so, we looked and sounded fantastic. Following the event, we changed into civilian attire and continued the festivities.

After hours of revelry, singing, and merrymaking in the bar, reminiscent of scenes from movies, we eventually bid farewell and embarked on our journey back home. The ride was just as joyous and lively as our celebration.

The day following our memorable performance, I set off to meet a friend in another part of Germany. As I gazed out the window during the journey, observing Germany's wintery landscape, the snowy scenery added to the Christmas ambiance. It truly felt like a white Christmas, with the cold temperatures further enhancing the holiday spirit.

My weekend in Vilseck, Germany, began on a picturesque note. I woke up to the sun streaming through a window, with a crisp chill in the air and a hint of frost. Spending the morning leisurely enjoying Christmas music and taking in the outdoor scenery was a delightful start to the day.

Later, I ventured out to explore the sights and attractions of the area.

Upon returning home from my visit, I made it a priority to decorate my living space, my home away from home.

The holiday experience in Europe was yet another first for me, and it was truly wonderful.

One memorable event was when my senior sergeant invited us over to her place to partake in some festive celebrations.

Experiencing a drink, never tried before!

I welcomed the beginning of my second year in the Army by watching fireworks and raising a glass of wine to a promising new year.

However, early in the new year, I received devastating news. I was tasked with the somber duty of helping to clear out my roommate's side of our room. She had visited the doctor late the previous year and was subsequently flown back to the states for specialized treatment due to a serious condition.

The news from the stateside treatment facility was heart-wrenching. Kris had been diagnosed with cancer, and it had progressed to an advanced stage. The news left me profoundly saddened.

While Kris was away at the treatment facility, she would often call me during breaks in her treatment. She needed someone to confide in, as she found it difficult to express her true feelings to her mother. That's where my friendly ear came in handy. During our conversations, Kris confirmed that she wouldn't be returning to Germany.

She was scheduled to undergo about eight to ten months of chemotherapy. Despite the challenges she faced, she sounded upbeat and optimistic. She mentioned that she would be stationed or living near the hospital and would ultimately be medically chaptered out of the Army.

"The only thing that really concerns me is losing my hair," Kris confided in me.

Kris's long, slightly curled, brown hair was always a prominent feature when she was out of uniform, and I could tell from her positive demeanor and humor that the prospect of losing it would be devastating for her.

While one friend was facing a challenging ordeal, another friend was about to experience the miracle of life as she went into labor.

At approximately 4:30 am, several days after my last conversation with Kris, I received a knock at my door. It was my very pregnant friend Lynne, informing me that she was in labor and ready to head to the hospital. I was designated as the driver for that important mission.

I found myself tasked with getting Lynne to the hospital, a duty that involved driving a vehicle with a standard transmission on the German autobahn in the darkness of the early morning.

We reached the hospital around 6 am, and Lynne remained in labor until approximately 11:37 when she gave birth. Leading up to the delivery, we spent the early hours offering support, assisting Lynne with her breathing, and providing comfort.

When the time came to move to the delivery room, we adorned ourselves with hospital gowns and entered. Initially, I had to wait outside as I wasn't technically the designated coach, just the driver. However, once the baby arrived, I hurried into the room with my camera in hand to capture the magical moments for Lynne and baby Lynne.

I was fortunate enough to capture all the precious moments of the arrival, from the baby's initial cleaning to the stitching and cleaning of Lynne, and finally, the transfer of baby Lynne into her arms, followed by the return to her initial room. The only part I missed was the actual delivery of the baby. However, I did have the breathtaking opportunity to cut the umbilical cord.

We all gathered around as the baby was measured, cleaned, and fed. We spent some time chatting with Lynne before leaving the hospital around 2 pm to head back home. The entire experience provided a welcomed distraction from the sadness we were feeling over Kris.

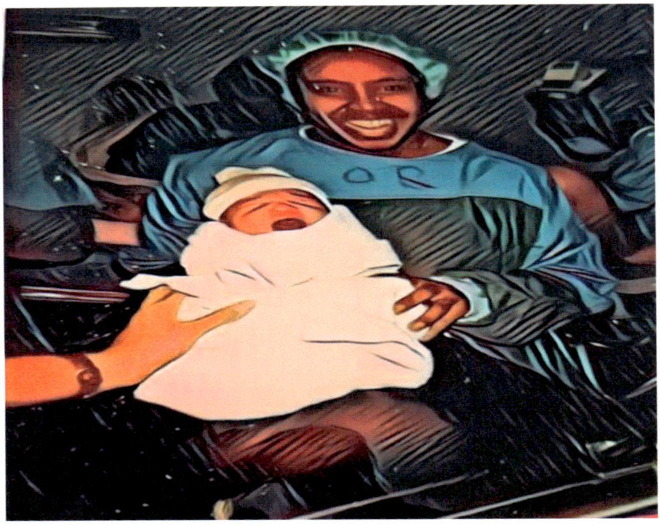

Punishment in the Army.

After the exhilarating experience of our friend's new baby's arrival, we were abruptly brought back down to reality as we returned to work, needing to prepare for our upcoming field exercise in the dead of winter.

During that early-year field exercise, I would have my first encounter with the military judicial system. However, before that momentous occasion, I endured grueling times in the field.

The first day started with a rocky beginning, as we had to rise super early to complete reconnaissance operating procedures. That involved arriving at the site before everyone else to set up the field area for their arrival. With so many tasks to accomplish, it quickly became overwhelming.

From camouflaging my vehicle solo to kitchen detail, that duty was truly the epitome of hell. The nightmare dragged on until about 10:30 pm, and I had to muster the energy to do it all over again the next morning.

The following night brought another hellish duty: guard duty from midnight to 1 am in the cold and rain. It felt like the coldest night of my life, and I was utterly miserable. After completing my duty, I returned to my warm tent, only to be rudely awakened by the First Lieutenant at 4:30 am, informing me that we had to return to the rear immediately. I was given a mere 15 minutes to prepare—a typical scenario in the military, characterized by last-minute orders and a constant sense of urgency. It seemed that nothing could ever be given sufficient time to be accomplished.

Upon our arrival at the unit, things initially went smoothly. However, a careless mistake on my part occurred when I left my weapon unsecured in the vehicle we returned in. Immediately upon our arrival, the vehicle was driven off to a class by a group of officers, who were unaware of the weapon's presence.

An unsecured weapon in the military is a grave offense, one that I knew could result in serious consequences. Though I was fully aware of the weapon's location, I was nonetheless filled with nervous anticipation regarding the punishment that awaited me.

In the moment, I immediately notified both the First Sergeant and the Commander of my mistake. The following two weeks were agonizing as I awaited my fate. Eventually, I learned that I would receive a summarized Article 15 with 14 days of laborious duty—a punishment considered to be the lowest possible for such an offense. Despite the gravity of the situation, I was overjoyed with the outcome, as it could have been far worse. I could have faced a loss of rank, financial penalties, and 45 days of laborious duty.

I was immensely grateful for the commander's leniency. His kindness spared me from any lasting blemish on my record that might have hindered future advancements. That act of generosity profoundly influenced my outlook and future actions as a leader.

The first Army school; road to promotion.

Two months after completing my judicial punishment, I was sent to my first Army education course, the Primary Leadership Development Course (PLDC). Over the course of four weeks, I immersed myself in learning essential leadership skills.

In a program tailored to prepare soldiers for leadership roles, I achieved a significant milestone by graduating on the Commandant's List. That list recognized students who maintained a 90% or better average on three written exams and received a first-time GO (passing grade) on all other areas of performance. For me, as a first-time participant in military education, that achievement was monumental. Among a company of 133 students, I graduated with the seventh highest score.

However, the joy of my graduation day was overshadowed by devastating news. Upon returning from school, my sergeant informed me and another soldier that my former roommate, Kris, had passed away from cancer. They had waited until after graduation to deliver the news, sparing us from sorrow during our training.

What should have been a day of celebration quickly turned into one of profound sadness. Kris was only 22 years old. A few days later, the unit held a memorial service in her honor, marking a somber occasion for all who knew her.

Closing Out First Two Years in Europe

You must see Europe.

After a period of grieving for Kris, I felt compelled to explore more of the European continent. There was no way I could be stationed there and not take advantage of the opportunity to visit nearby countries. So, I embarked on a journey to travel throughout the continent.

First stop Italy.

Si-yes
Chow-bye
Saldi- sale
Graci-thank you
Amore'-love

The Italian Lire

Florence-Florentina
Venice-Venezia
Verona-home of Romeo & Juliette

Canonica Canal

Welcome to Switzerland

But what's a trip to Europe without visiting Berlin? I couldn't miss the chance to see the Wall before it was painted.

AUF DER MAUER-on the wall

Ja-yes
Danke-thank you
Auf Weisben-bye

Concert Hall – Schauspielhaus Berlin

The Rhine River

Paris was a must!

The Eiffel Tower

While in France, I made sure to visit iconic landmarks such as the Notre-Dame Cathedral, the Palace of Versailles to admire its remarkable works of art, and the Louvre Museum.

My vacation in Europe was unforgettable, but soon it was time to return to work and resume my duties as a soldier, preparing for the upcoming major training exercise.

The Return of Forces to Germany (REFORGER) was a massive month-long exercise involving units from all over the world. It was meticulously planned and served as an opportunity for forces to collaborate and train together.

Living in a field environment for an extended period presented various challenges. We had to get creative with our living conditions, using meals-ready-to-eat (MRE) boxes as makeshift drawers for clothing. As the only female in my platoon during field exercises, we improvised privacy using our ponchos.

REFORGER, a month-long field training exercise!

After the conclusion of REFORGER, we resumed our routine duties back at the unit. A few months later, I would bring my first tour in Europe to an end.

Years Later…Another European Tour

South-Central Germany-Part II

Back in Europe!

It would be a decade before I returned to the European continent for another tour of duty. The second tour would be markedly different from the first. As an officer, I was bestowed with a host of privileges and responsibilities.

My new unit embraced the motto SEMPER PARATUS, always prepared. Situated on a small Kaserne nestled between two locales, it would serve as my home for the next three years.

During my second European tour of duty, I encountered a plethora of new experiences and celebrated more holidays on the enchanting European continent.

One of the highlights was a visit to the ChristKindle Market in Nurnberg, Germany, where I immersed myself in the dazzling lights and rich cultural traditions.

Additionally, I had the privilege of attending my first Air Defense Artillery St. Barbara Ball in Wurzburg, Germany. St. Barbara, the patron saint of artillery, was honored during that event, and those who had made significant contributions to the air defense artillery branch were bestowed with the prestigious medal.

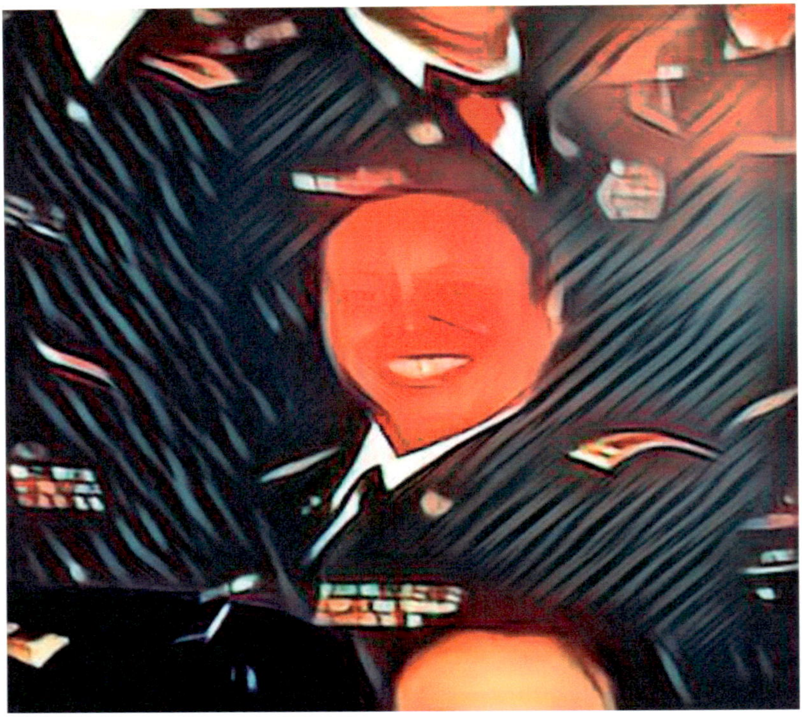

Amidst my duties and adventures during that tour of duty, I achieved a significant personal milestone: I graduated with my PhD in Education.

Similar to my first tour of duty in Europe, air defenders found themselves involved in conflict once again. While Operation Desert Storm marked the first conflict, my second tour coincided with Operation Iraqi Freedom.

Upon the unit's return to Europe following the completion of the mission in Iraq, there was a noticeable change in personnel. The experiences of the mission had left a profound impact on individuals. Recognizing the need for rest and reintegration, everyone was provided with much-needed down time and reintegration time.

Several months after the unit returned from Iraq, I made another permanent change of station move and moved back stateside. My combat tour in Iraq marked the end of my second European tour, and I would not return to the region for any further military assignments.

Before assuming duties at my upcoming stateside duty station, I completed my Warrant Officer Advance Course, preparing me for another promotion.

Made in the USA
Las Vegas, NV
07 November 2026

57669922R00031